FRESHWATER AQUARIUMS:

PROPERLY SET UP YOUR TANK AND LEARN HOW TO MAKE YOUR FISH GROW.

Table of Contents

INTRODUCTION 7

CHAPTER 1. GETTING STARTED WITH THE AQUARIUM 10

    TANK 11
    TANK STAND 11
    FILTER 11
    HEATER 12
    THERMOMETER 12
    LIGHT 12
    PLANTS 13
    DECORATIONS 13
    SIPHON 13
    BUCKET OR HOSE 13
    FISH NET 14
    TEST KIT 14
    DECHLORINATOR 14
    AMMONIA 14
    SEEDING MATERIAL FOR BACTERIA 14

CHAPTER 2. HOW TO ASSEMBLE A FRESHWATER AQUARIUM 16

    SETTING UP YOUR FRESHWATER AQUARIUM 16
        Step 1 – Where to Place Your Aquarium 17
        Step 2 – Leveling Your Aquarium 18
        Step 3 – Preparing the Aquarium 19
        Step 4 – Adding a Background 19
        Step 5 – Adding Gravel 20

Step 6 – Adding Filtration 20
Step 7 – Attaching a Heater 21
Step 8 – Decorating Your Aquarium 22
Step 9 – Adding Water to Your Aquarium 22
Step 10 – Install a Thermometer 24
Step 11 – Start the Filter 24
Step 12 – Plug in the Heater 24
Step 13 – Topping Your Aquarium 25
Step 14 – The Final Step: Water Conditioner 25

CHAPTER 3. FISH AND HOW TO CARE FOR THEM 26

    FEEDING 26
    GARDENING 27
    WATER CHANGE 28
    FILTER CLEANING 30
    PANE CLEANING 31
    DISEASES 32

CHAPTER 4. WHAT KIND OF FISH TO BUY 34

    BUYING A HEALTHY FISH 34
    FISH QUANTITY 35
    COMMUNITY FISH AND AGGRESSIVE FISH 37
    FISH COMPATIBILITY 39

CHAPTER 5. DIFFERENT SPECIES 40

    DIFFERENT TYPES OF AQUARIUMS 41
    TYPES OF FISH 42

CHAPTER 6. HOW TO FEED DIFFERENT FISH 45

SIZE 45

VARIETY 46

AMOUNT 46

FREQUENCY 46

VARIETY IN FEEDING 46

WHEN TO STOP FEEDING 47

FRESH VS. FROZEN 48

FISH WEIGHT 48

CHAPTER 7. HOW TO MAINTAIN HEALTHY FISH 49

CHAPTER 8. DISEASES AND TREATMENTS 52

FISH DISEASES: CAUSES AND DIAGNOSIS 53

CHAPTER 9. WATER CHEMICALS AND PLANTS 56

DIFFERENT TYPES OF WATER 56

Differences between fresh and saltwater *56*

Water *57*

OTHER CHEMICALS 58

HOW TO CHOSE PLANTS 59

Set up your aquarium plants *60*

Use water conditioners *60*

CHAPTER 10. THE NITROGEN CYCLE OF YOUR AQUARIUM 61

DIFFERENT STAGES IN NITROGEN CYCLE 62

CHAPTER 12. TAKING CARE OF FISH 65

DIET & NUTRITION 65

Flakes *65*

    Freeze-Dried Foods 66

    Frozen Food 66

    Live Food 67

  QUANTITY AND SCHEDULE 68

  COMMON BEGINNER CARE MISTAKES 69

    The Fish Will Grow to the Tank Size 69

    Over or Underfeeding 69

    Improper Medications 70

    Not Cycling the Tank Properly 70

    Overstocking 70

    Using Small Tanks or Bowls 71

    Omitting Equipment 71

CHAPTER 13. TECHNOLOGY AND EQUIPMENT FOR FRESHWATER AQUARIUMS 73

  BASIC TECHNOLOGY AND EQUIPMENT 73

  AQUARIUM STANDS 74

  CHILLERS 78

CHAPTER 14. THE CORRECT STAND FOR YOUR AQUARIUM 81

  WHAT TO THINK ABOUT WHEN CONSIDERING AQUARIUM STANDS 81

    The size, weight, and shape of the aquarium 81

    Understand the different materials used for a stand 82

    Space for equipment 85

    Opening and height of your stand 85

    The desired, overall look of your aquarium 86

CHAPTER 15. SAFETY 87

    Drip Loops 87

    Surge Protectors 88

    GFCI's 88

    Grounding Probes 89

    Level Floor 90

CHAPTER 16. PROBLEM SOLVER CHART 92

    Stressed fish after water change 92

    Gasping for air at surface by fish 93

    Slow fish activity 93

    Fins rotting and blood on edges 94

    White dots (ich) or velvet coatings that are on the fish's body and fins 94

    Tails clamp 94

    Belly becomes flat or sunken 95

    Fish head points up, spins and uncontrolled swimming ability 95

    Snails migrate and stays on top or pull into their shells for long periods 95

    Snail shell becomes thin and brittle 96

    Cloudy water or green water 96

    Algae growth is excessive 97

    Filter not processing ammonia 97

CONCLUSION 98

Introduction

A freshwater aquarium is a closed environment where you keep water with live plants and animals inside it. It is like a miniature man-made river and the term "freshwater" means it has no salt in it. A fish tank that contains salt water would be called a marine tank, not a freshwater one.

An aquarium can be a bowl with a lid, a glass tank or plastic or acrylic box. The water, the fish and all the other living things inside need to be protected from dust and insects. For this reason it is always best to cover an aquarium with a lid. The water itself needs to be protected from the air, so that the fish cannot taste or smell anything from outside that may make them sick. Because of this you should always provide your aquarium with some kind of filter system, which constantly moves water through it. This way the water does not become polluted by any dust or smells from outside that may harm your fish.

A filter system can consist of an air pump and an air stone which will bubble the water. This removes any airborne pollutants such as dust and also keeps the oxygen level high enough for your fish to live.

The water needs to have other things in it besides fish, such as plants or rocks so that your fish feel safe from larger predators. There are many different types of aquarium plants that you can grow inside your tank. Some are real and some are fake, but they all help keep the water clean and oxygen levels high enough for the fish to live in a healthy way.

Aquariums are places where you can see some very interesting creatures up close and personal. Your fish will come to know you since you will be one of the few things they can see. People who like to take care of things and who like to keep living things happy are good candidates for owning a freshwater aquarium.

There are many different species of fish that you can keep in your aquarium. When first starting out you should start with some very small fish such as freshwater shrimp, livebearers and minnows. When these small fish get bigger they may eat the smaller fish or they may grow too large to be kept in a small tank.

One thing that you need to know about fish is that they are all very different from each other, which is why there are so many different kinds of them. The two largest groups of them are the sharks and the rays. Rays are actually bony fishes and have scales on their back, while sharks belong to the cartilagenous fishes which have skeletons made entirely of cartilage.

As far as types of fish go all the different colors are no big deal. Some fish have very vivid and bright patterns, while others have plain dull colors. This is just how they chose to live their lives.

There are many different fish that you can keep in a freshwater aquarium and even more that you should never keep in it. Many of these harmful fish are very strange looking creatures which many people do not realize they can house in a tank. The most venomous fish of all are snakes, but there are many other odd fishes that you should never keep either because they may be poisonous or because they will simply outgrow your tank at an unnatural time, becoming too big for it to contain them well.

Chapter 1. Getting started with the aquarium

Starting a freshwater aquarium as a beginner can seem a bit overwhelming at first. There is so much equipment that you need to get, and so many mysterious processes involving water testing and adding just a bit of this or that chemical.

Owning and maintaining an aquarium is a big responsibility to be sure. But if you compare it to owning most other pets (cats and dogs, for example), it's much less work. Most of the intensive effortcomes at the beginning, and once you've got your aquarium up and running, you'll only have to do minor maintenance tasks from day to day.

It's easiest to understand all of the steps in setting up your aquarium if you know a bit about how an aquarium works.

Fish live, breathe, eat and excrete waste all in the same water. You need to provide them with oxygen to breathe, food to eat, and some way of processing the toxins in their waste.

The filter agitates the surface of the water so that it gets re-oxygenated. It also keeps the water clear by filtering out particles of waste. What's less obvious is the role that bacteria play in breaking down the waste into harmless substances. These bacteria live in the filter and on surfaces inside the aquarium, and you need to look after them, too.

To get started, you can buy a complete aquarium kit with everything you need, but it may be cheaper to buy all the elements yourself, and you will also have more control over the end product.

So here's what you'll need:

Tank

You may think that starting with a small tank would be easier, but in fact smaller aquariums are harder to look after. It's easier to control the water quality in a tank that's at least 20 gallons. Think about the kind of fish that you want, and research their space requirements. Also take the weight of the tank into consideration, and make sure your floor can support it.

Tank Stand

This has to be able to support the weight of the filled tank. It's best to buy a stand that isspecially designed for aquariums, or have one custom-built. Repurposing an old piece of furniture can get you into trouble if it's not strong enough.

Filter

There are several different types of filter available.You want a strong filter, so look for one that's rated for the aquarium size that you have, or even bigger.It should hang on the back or just inside your tank, and it may have compartments for different filtering materials. Charcoal is not usually necessary (unless you've just medicated your fish for a disease) but it also does no harm.

Heater

There are some fish that do well in colder water, but a heater will give you a lot more options when it comes to stocking your aquarium. Look for a heater that's rated for the size of your tank and make sure it has an adjustable thermostat.

Thermometer

A mercury thermometer that sits inside the tank is best. The thermostat on your heater alone is not enough to give you an accurate reading on how warm the water is.

Substrate (Gravel or Sand)

Both of these are fine for the bottom of your aquarium. Gravel is a little easier to prepare and look after, but sand creates a habitat for certain burrowing fish that you might be interested in stocking.

Light

These are often included when you buy a tank. If you're going to have live plants in your aquarium, do some research and make sure the light is bright enough for them.

Plants

These can be real or artificial. Real plants are a little more work, but they also improve the water quality and give fry (newborn baby fish) a place to hide. Silk and plastic plants are also fine, and many of them look surprisingly realistic. Just make sure they're designed specifically to be used in freshwater aquariums.

Decorations

Here's where you can let your imagination run wild! You can go for a natural look, or you can create a theme park for your fish, complete with medieval castles, steampunk ornaments, or even Spongebob's house. As long as the ornaments are all specifically made for freshwater aquariums, and as long as they give fish some places to hide away, it's entirely a matter of personal taste.

Siphon

This is a simple tube, often with a wide attachment at one end that allows you to remove water from the tank and "vacuum" the substrate to remove impurities.

Bucket or Hose

Your tank may be close enough to a tap to rig up a hose attachment. Otherwise, you can use a bucket to fill the tank and do water changes.

Fish Net

You'll need this to add or remove fish.

Test Kit

You'll need to test the water for various chemicals, including ammonia, nitrites and nitrates. Don't use dip strips; invest in a liquid drop test kit that will give you more accurate results.

Dechlorinator

This comes in a liquid form. Buy one that gets rid of both chlorine and chloramines. Chlorine is added to our drinking water to kill harmful bacteria, but it will also kill the friendly bacteria in your aquarium, making it unable to support fish life.

Ammonia

Make sure you buy 100 percent pure ammonia, with no surfactants or cleaning agents added. You will use this when you are cycling your tank, or preparing it for the fish to live in.

Seeding Material for Bacteria

You need a way to introduce bacteria into your new tank. The best way is to use some filter media from an established, healthy freshwater aquarium in your own filter. You can also borrow some gravel or decorations from an established aquarium. Keep these items wet at all times before installing them in your own aquarium.

If you don't have any friends with freshwater aquariums, you can also buy bio-filters that are seeded with friendly bacteria. There are also liquid bacteria supplements, but there is a lot of disagreement about whether these actually work.

Chapter 2. How to Assemble a Freshwater Aquarium

This is the fun part: the actual aquarium assembly! And we are one step closer to adding our aquatic friends. First, look at the equipment to build our freshwater aquarium. Next, check out some optional products. Finally, find a list of necessary test kits.

Setting Up Your Freshwater Aquarium

Equipment:

- Aquarium
- Aquarium Stand

- Top and Light
- Filter
- Heater
- Thermometer
- Power Strip
- Background
- Gravel
- Decorations
- Water Conditioner
- Bacteria Starter
- Fish Food
- Fish Net
- Bucket
- Siphon Vacuum

Step 1 – Where to Place Your Aquarium

Even small aquariums are extremely heavy when full, with just the water weighing about 8.3 pounds per gallon – plus the gravel and the aquarium itself. A fully set up 38-gallon aquarium will weigh more than 355 pounds! Try to settle on a permanent spot prior to set up. If you are using an existing piece of furniture in place of a stand manufactured to hold an aquarium, be sure that it will hold the weight.

Additional care should be given to the placement of tanks larger than about 150 gallons, as the floor underneath may not be suitable to hold such weight.

Keep in mind that aquariums receiving direct sunlight will grow an abundance of algae. While algae are not harmful to fish, it will mean more cleaning to do later. Avoid putting your aquarium near a window.

Step 2 – Leveling Your Aquarium

Whether you have purchased an aquarium stand or are using existing furniture, make sure the base is level before you begin filling your aquarium. Once the aquarium is filled, it will be obvious if the tank is not level. This is not a good step to skip!

Use a level to check front to back, side to side, and from corner to corner. Use shims under the stand at the floor until the tank is level.

When installing the aquarium on carpet, know that wood or metal strips are often used to secure carpet to the floor and to keep it flat. These strips are installed close walls. If the aquarium is placed on a strip, it will raise up the back of the aquarium. If possible, space the aquarium from the wall in front of the strip.

Once shims are placed under the stand and the tank reads as level, you are ready to move on.

Step 3 – Preparing the Aquarium

Using a soft cloth or damp paper towel, wipe down the inside the of the aquarium to remove dirt and dust. Never use soap or household cleaning agents in or on your aquarium.

Step 4 – Adding a Background

The background is attached to the outside back of the aquarium to hide the filter and electrical cords. It adds depth to your tank and can create a pleasing setting.

Clean and dry the back of the aquarium to prepare it for the background. Cut the background to the exact length of the aquarium. Using clear tape, attach the right and left sides to your aquarium. Next, attach the top edge of the background. Run a piece of tape the entire length of the aquarium to keep drips and splashes from getting between the background and the aquarium.

If a solid color is desired – typically blue or black – spray paint can be used to coat the back of the tank before the tank is put in place. Choose the desired color in waterproof or water-resistant matte finish. Only use spray paint outdoors or in a well-ventilated area. Use painters' tape and plastic to protect from unwanted overspray. Spray the tank upside down so that no paint gets into the tank. Allow the paint to dry fully before moving the tank. A razorblade can be used with caution to scrape off painted backgrounds.

In general, dark colored gravel and backgrounds will contrast more with fish to accentuate the colors of community fish and make them pop.

Step 5 – Adding Gravel

If the gravel you chose looks dusty or dirty, rinse it in a large bucket, 10 to 15 pounds at a time. Run clean tap water through it, stirring gently with your hand, and drain the bucket to rinse away dust, debris, and color flakes. A kitchen colander also works very well.

Add enough gravel to create a layer between ½ inch and 1 inch thick for a fish-only aquarium. For most aquariums, this will be a little less than a pound per gallon of tank volume. An aquarium with live plants needs smaller gravel particles and a deeper base of 3 to 4 inches.

In metric units: 1.25-2.5cm of gravel for fish-only; 7.5-10cm for planted aquariums.

Step 6 – Adding Filtration

There are several types of filters that can be used with your aquarium.

Most tops have cut-outs at the back to accommodate the filter. Glass lids come with plastic back strips that can be trimmed to fit the filter. Position a power filter or tubing for a canister filter to align with the cut-outs or where the back strip will be trimmed. Follow the manufacturer's instructions for setting up the filter appropriately. If the intake tube can be adjusted, the intake should be set to about two-thirds of the way from the top of the aquarium. For optimal flow, the intake and return for a canister filter should be in opposite back corners.

Do not start the filter at this time. Without water the filter pump will be damaged.

Step 7 – Attaching a Heater

Nearly all modern heaters are fully submersible, and most have pre-calibrated thermostats built in. Follow the manufacturer's instructions to calibrate the heater if necessary. Secure the heater to the glass with the included suction cups. The ideal location for the heater is near the filter's return so that the water flow will disperse the warm water throughout the aquarium.

Large tanks may require two heaters on opposite sides of the tank to adequately heat the large volume without creating pockets of warm water and cold water.

Do not plug in the heater at this time. Without water the heater will quickly overheat and be damaged.

Step 8 – Decorating Your Aquarium

Rinse all rocks, artificial plants, and other décor. Smaller, shorter plants should be placed in the front of the aquarium, with larger, taller ones toward the back.

First, if you are using driftwood in your aquarium, place it a little towards the back. The driftwood can be the focal point with decorative rocks and plants as accents. Rocks and driftwood are used to create levels, like caves, in your aquarium for fish to swim through.

Experiment with positioning the décor in different places. Be creative with this enjoyable step! There are many ways to aquascape your aquarium; the most important part is that you are happy with the way it looks. Remember, you can always make changes.

Step 9 – Adding Water to Your Aquarium

For smaller aquariums a 5-gallon bucket filled at the kitchen faucet or bath tub is all you will need to fill the tank. Try to adjust the water temperature to between 80°F and 82°F (~27°C) directly at the faucet.

For larger aquariums, carrying water back and forth may take too much time and energy; unless you're looking for a workout, you may want to use a garden hose. If you do, let water run through the hose for several minutes before filling your aquarium to flush stagnant water from the hose.

When using a garden hose, you can adjust the temperature by adding hot water, but only after the tank is partially full. Hot water directly on cold glass can cause the glass to crack.

There are some useful maintenance kits available which you can use to fill your aquarium and that attach directly to a kitchen faucet with 25 to 50 feet of tubing. These kits can also be used to vacuum the aquarium directly into the sink.

Whatever method you use to fill your aquarium, it is important to avoid stirring up the gravel while pouring in the water. You just took time to arrange your decorations. Dumping water carelessly onto the gravel will mess up your work! It may also cause a cloudy tank, especially if the gravel was not sufficiently rinsed. Pour the water (or point the hose) onto a large rock or decoration. You can also place a dinner plate in the tank temporarily and pour water over that.

Fill the tank to over the bottom of the top frame, or about ¾ of an inch below the top of the aquarium.

Step 10 – Install a Thermometer

Attach a hanging thermometer to the aquarium on the side opposite the heater, about three inches from the top of the aquarium. Keeping a thermometer in the aquarium allows you to confirm that the heater is working properly and that in hot summer months the aquarium is not too warm.

Step 11 – Start the Filter

Now it's time to prime the filter. Using a cup or pitcher, begin filling the filter manually before plugging it in. If the filter doesn't start running, continue filling it manually until the pump fully engages. If your filter uses cartridges or carbon, rinse them before use.

Your filter should turn over the water in the aquarium between 5 to 8 times per hour.

Step 12 – Plug in the Heater

Allow the heater to adjust to the temperature of the aquarium before plugging it in to prevent the glass from cracking. The heater should be set between 80°F to 82°F (~27°C) for the first four to six weeks until the biological filter is established.

Step 13 – Topping Your Aquarium

Remove the tabs or trim the plastic back strip to fit the location of your filter. Place the hood or glass tops on the aquarium and plug in your light.

For planted aquariums, the light should be on between 10 to 12 hours per day. With no live plants, run the light for no more than 8 hours per day. When the photoperiod is too long, your aquarium water may eventually turn an unsightly green or red from algae.

Standard light fixtures that come with most starter kits are not suitable for growing plants. Low-light plants like java fern may do well enough, but most require a more powerful light to thrive. If you have a standard fixture, artificial plants are best.

Step 14 – The Final Step: Water Conditioner

Add a good water conditioner to neutralize harmful chlorine and chloramines. Follow the manufacturer's instructions for the dosage. Even freshwater fish require a small amount of salt, which should be added at this time unless you have live plants.

Chapter 3. Fish and how to care for them

The fish and plants are doing well. Everything has developed magnificently. You enjoy looking at your little underwater world. But this alone is not enough. To keep it that way, care is necessary. Some things are done quickly, others take longer. And perhaps there will also be one or two things that you won't like so much, because you might have imagined it quite differently. In such moments just think about the beautiful aquarium and how great everything has developed. You see, everything is not so bad.

Feeding

People like to do this and it is always exciting to watch. But here applies, less is more. You should find a good compromise, so that on the one hand all animals are fed and on the other hand no food is left lying around. The food that is left lying around pollutes the water and can be the cause of poor water quality. You remember the bacteria that convert waste into nitrite? Therefore it is better to feed sparingly. Also, you will have more time to watch.

You should know your inhabitants and accordingly have suitable food in stock. There are certainly fish that need to be fed very specifically. Not only the type of food, but also the way of feeding. Therefore inform absolutely before purchase of the animals exactly about the conditions! With these mentioned animals usually the usual and widespread flake food and food tablets for the bottom dwellers are sufficient. However, even here you have to take into account whether it is a carnivore or vegetarian and choose the appropriate product accordingly. But we humans also do not want to eat the same thing every day. Therefore, it can also be worthwhile to deal more extensively with the topic of food and to bring variety to the menu. Your animals will thank you with health and more beautiful colors. Beside flake food there is also frozen food in the trade, which is also easy to feed. You can make the menu even more interesting by growing the fish food yourself. In the form of microorganisms, such as water fleas or Artemia. However, you can deal with this topic later. Ready-made food is sufficient for the beginning.

Gardening

This is likely to be the maintenance work that is the most time consuming. Again, of course, it depends a lot on the amount of plants you have. But if your plants are growing well, you can expect to have to cut back some plants every two weeks. If the tank is too overgrown, it usually doesn't look nice either and the fish still want some swimming space too. So get out the sharp scissors and go. But how and where?

The easiest way is with stem plants. You can simply cut them off at the top and you can even stick the head back into the ground. After a short time it will grow roots again and you will have a new plant. If the plant is no longer so beautiful at the bottom, you can also replace the unsightly part in this way so. But that then not with all plants at the same time, that could bring your stable biotope again out of balance.

With the Amazon sword plant it can happen that it grows too high or also becomes too powerful in the width and takes away space or light from other smaller plants. Then you should cut off a few leaves as close to the ground as possible. You can also propagate this plant. It quite often forms shoots on which small daughter plants are formed. You can plant these as soon as roots have grown.

Mosses or grass-like plants spread all by themselves and form a dense carpet over time. But this usually requires a lot of light for them to grow compactly and regular pruning, and you can also replant the cut plantlets in the ground. However, this can be a test of patience, since the small plants hardly find a foothold as long as there are no roots.

Water change

Our biotope is well run in and yet it remains an artificially created biotope, which does not exist in the wild. As well as the metabolic cycle works, it comes in most aquariums sooner or later to the point that some substances can not be degraded. And then they remain in the aquarium and the time comes when the biotope is no longer stable or the animals are not doing so well and may even die. To prevent this, the pollutants must be regularly removed from the aquarium. The easiest way to do this is to change the water. The emphasis here is clearly on regular. 30% of the water volume every two weeks is sufficient. This is not much and, depending on the size of the aquarium, can be done quite quickly. Ideally, you can combine this directly with the bottom cleaning and at the same time the next maintenance task is done. For small and medium aquariums, you simply take a short hose, one end you hold in the bucket, which is slightly lower than the aquarium. The other end goes into the aquarium. Now suck once briefly and vigorously on the hose end in the bucket and the water runs from the aquarium into the bucket. Almost all by itself. Just make sure that no fish or other inhabitants are transported into the bucket. Otherwise they have to get out of the bucket and into the aquarium again, which can be tedious with the landing net. If one has the 30% out, they must also again purely. The easiest way to do this is with the bucket. Get fresh water from the tap and pour it carefully into the aquarium. But if the aquarium is larger, then the bucket can quickly become laborious, because you may have to run 10 times or even more. Here you can proceed according to the same principle, but with a long water hose directly to the bathroom. However, due to the length of the hose, the flow rate may be slower and the water change alone may be longer. But you save yourself some running around and lugging buckets.

As already mentioned, it is quite good to do a regular water change with not too much water. But sometimes you have to deviate from the rule. This is necessary if the nitrate value has suddenly risen or other pollutants contaminate the water. This puts our inhabitants in danger and a large water change is urgently necessary. In this case even up to 90%. Until the water values are stable again in the safe range, this may even be necessary several times a day. But this is not the rule and should not scare you off! A small tip! You do not need to throw away the aquarium water. As watering water, for indoor or garden plants, it is perfectly suitable.

Filter cleaning

In my opinion, the filter is the only device that is a bit more complicated to maintain. And here, the smaller the filter and the more stock in the aquarium, the more often you have to clean it. But that's not witchcraft either. A small rule in advance, you should not clean the filter on the same day as the water change is done. This could cause too many bacteria to disappear from the aquarium at once, and we need them.

The small internal filters are usually only foam sponges or cartridges. Depending on the manufacturer, you can rinse them briefly under lukewarm running water or simply replace them with a new one. The larger external filters are more complex. However, even here everything is rinsed only with lukewarm water. Here it is only important that you assemble everything in the reverse order to how you disassembled it. When to clean the filter? Some manufacturers specify a cleaning cycle. However, I keep it very simple. If the flow rate has decreased significantly or the water is no longer filtered clearly, then the filter should be cleaned. By the way, the hoses of the external filter do not have to be cleaned every time. Deposits also settle here, but experience has shown that it takes much longer until the flow is prevented.

Pane cleaning

The frequency of glass cleaning depends more on the growth of algae. If there is little algae in the aquarium, then there is also less coating on the panes and it must be ensured less often for clear view.

If it should be necessary, there are different possibilities. Very classical are magnetic cleaners. A part equipped with smooth felt or fabric remains on the outside and a part equipped with rough material is placed on the inside of the pane in the aquarium. If you now move the outer part, the counterpart on the inside also does it and eliminates the algae. You can also leave this solution in the aquarium and quickly „go over" it if necessary. But you should avoid getting into the bottom at all costs. If a stone or grain of sand gets between the magnets, scratches can no longer be avoided! By the way, floating magnets prevent wet hands when fishing out again.

As alternatives, there are countless sponges on offer or disc cleaners with razor blades. The latter are very effective and are available on a long rod, so that you do not have to use your hands in the water. In the variant with the razor blades, however, it is important to work carefully so that no scratches come into the glass and especially the silicone seams are not injured. But as long as you slowly put the razor blade always straight on, nothing happens.

From the outside, you can also clean the windows from time to time.

Diseases

Unfortunately, fish can also become ill. There it goes to the animals quite similarly as us humans. But honestly, the diseases of humans are better researched and also better treatable. Especially with small fishes one usually has little time to do anything at all. It is also difficult to always find a reason why a fish died. In any case you should measure the water values in this case. So you can at least exclude that a water value has slipped into a dangerous range. And other animals are also in danger.

However, if several fish show the same symptom, or even just a single one, then the best thing to do is to set up a small sick bay and quarantine the sick candidates for the time being. This means, of course, that you have to set up a separate aquarium. It doesn't have to be that big and it doesn't have to be overgrown. But you should have thought in advance where you can set up such a sick bay for a short time. It is important to note that you fill the infirmary with water from the main aquarium. The small tanks are usually operated with small internal filters. This can be run in the large tank a few hours before. Then a few bacteria have already settled there. You can do without plants and substrate for the time being. In this tank you can now take a close look at the patient and then consult specialist literature. You can also try a specialized store. But then please in a real one. Nothing against hardware stores, but usually only salesmen are employed there. In a specialized store the chance is higher to meet an experienced aquarist. They have their own experience with one or the other diseases and medications. In the accessories there is a whole range of medications for various diseases. Most of the treatments require isolation of the patients and it is highly advisable to stick to the given dosage. And here, too, then usually helps only patience and hope.

Chapter 4. What kind of fish to buy

Buying a healthy fish

Searching for the perfect pet to occupy your tank must be done in a thorough planning. Of course you want something that is worth your money, time and effort.

The first step into buying a healthy fish is making sure that you are transacting in a reputable pet shop. First of all, do a research about fish tanks and ask the staff some questions related to what you have learned.

If they know what they are talking about, then, you are dealing with credible people. Observe around the store.

Look closely into the displayed tanks to see if there are dead fishes floating in there. A good store would not allow dead fishes into the displays, especially if it is co-inhabiting with live ones.

Observe if there are fishes that have small white spots or other symptoms of sickness. Do not buy a fish that comes from a tank with sick inhabitants.

Tinted water may also be a sign that the tank houses a sick fish. To ensure that the fish is healthy, it must have the following characteristics: clear eyes, active and alert movement but not too hyperactive, steady breathing, fins must be in good shape and condition, must not be bloated, must not be slimy and must have a healthy color.

More importantly, it must look clean and not have any blemishes in the body as these can be signs of stress or bacterial infection.

Once you have decided and picked a healthy fish to buy, ask the staff to add more water in the fish bag if you are going to drive for 15 minutes or more.

Fish Quantity

It is safe to say that more fishes in the tank provide more entertainment. However, you should always remember that overpopulation would possibly result to a disaster.

In determining the number of fishes to house in your aquarium, the basic ratio to follow is an inch of fish is to one gallon of water. While this rule might give a rough estimate for a good start, it must also be considered that fishes would grow as time goes by.

Therefore, in using this rule, calculate using the typical adult size of a fish. Consider the shape of the fish as well.

It is wrong to assume that the tank's size is equivalent to the volume of water it can hold. Example, a five-gallon tank filled with plants, gravel and other decors does not really hold 5 gallons of water.

Actually, there must be a 10-15 percent difference between the tank size and water volume.

Another way to determine the number of fishes that a tank can hold is by calculating the tank's surface area. To calculate the surface area of a tank, multiply its length by its width.

This method uses a ratio of 1 inch of fish is to 12 square inches of the tank. This ratio applies only to slimmer fishes.

For wide-bodied fishes, the ratio must be an inch of fish per 20 square inches. This rule would work better than the one-inch rule if the aquarium is a non-standard size.

Another method to determine an aquarium's substantial population is by calculating the ratio of fish weight by the water volume. Typically, the ratio must be one gram of fish is to 4 liters of water.

Community fish and aggressive fish

One of the most exciting parts of setting up an aquarium is choosing the fishes to keep. There are wide choices as there are literally a thousand species of fishes.

However, choosing the inhabitants to shelter in your aquarium is not as easy as picking up whatever you fancy. There are fishes that go peacefully with everyone else while there are also some that come as a threat to others. Freshwater species are categorized based on these temperaments.

Community fishes are those that are considered peaceful. They are harmonious in nature and can be mixed well with almost every species.

They behave well when grouped with other fishes belonging to the same species. However, there are some community fishes that can become aggressive when kept alone; such as tiger barbs and serape tetras.

Although they may be aggressive when alone or kept in pairs, they can still be peaceful when kept in large groups. It is recommended to shelter them in groups of 4-6 with their other co-species.

Examples of community fishes are danios, guppies, platies, swordtails, mollies and majority of tetras.

The next category is the semi-aggressive type. These are peaceful fishes that can still display aggressiveness under some circumstances.

If you want to keep this type of fish, make sure that there is only one semi-aggressive male in the community. Also, be sure to provide more hiding places inside the aquarium. This way the smaller fishes can protect themselves in case the semi-aggressive displays threat and aggression.

If you want more than one semi-aggressive fishes, make sure that your tank is large enough because they can be territorial. They will get along fine as long as there is a spot that they can consider as their territory.

Examples of semi-aggressive fishes are eels, loaches, barbs and gouramis.

Aggressive fishes, on the other hand, have wild temperaments. They are hardly compatible with other species.

They exude a strong territorial aggressiveness especially if a male fish is mixed with another male fish. Fishes belonging in this category may be mixed with other co-species but are still better kept alone.

They also tend to be aggressive on fishes that look similar to them.

Housing aggressive fishes may require more expertise that is why they are not recommended for newbie aquarists. Examples of aggressive fishes are Plecostomus, large catfish, and most cichlids like African and Oscar.

Fish Compatibility

In order for your little freshwater community to stay serene, the inhabitants must be compatible with each other; they must co-exist peacefully. As previously mentioned, there are species that can go along well with any other species while there are some that cannot.

Here are some specific examples of popular tropical fishes that are and are not compatible with each other.

Guppies can be grouped with other small fishes like Platties and Mollies. They prefer alkaline water.

Therefore, these peaceful fishes are not advisable to be mixed with others that prefer acidic and soft water.

Barbs and Cichlids should not be mixed together. The latter have long fins and the former is a known fin nipper.

Guppies should also not be mixed with fin nippers such as gouramis. If keeping guppies, follow the ratio of two females per one male to help maintain peace in the community.

An angelfish may also be meant for schooling but with some exceptions. It cannot be mixed with neon tetras and guppies as it may show aggression around these species.

Female bettas are good for schooling but take caution among other species that may nip at their fins. Male bettas, on the other hand, must be the only betta in the aquarium. Better yet, it must be sheltered in a separate tank.

Chapter 5. Different Species

Different species in aquarium. There are many different types of fish to choose from when setting up an aquarium. From tangs and butterflyfish to gobies and gouramis, there are plenty of fish variety. Before we get to the different types of fish in aquarium, let's talk about what is a good tank size for a single betta fish.

Different types of aquariums

There are many different types of aquariums that can be found at most local pet stores. It would not be possible for me to mention them all or even list them in any sort of order as each one will have its own positives and negatives that should be considered before purchase is made.

The most popular aquarium size is the 15-gallon. This can be used for both tropical and coldwater fish. Anyone on a limited budget will find that 15-gallon aquariums are usually the cheapest and will easily fit into any room in your home or apartment.

For many people, the 10-gallon aquarium is more than enough to keep one fish (such as a betta) happy and healthy. This size aquarium can be used for both tropical and coldwater fish, but should not be used for any type of reef tank unless you want to have a very small collection of invertebrates.

Most 5-gallon tanks are smaller versions of 10-gallon tanks. They can be used for both tropical and coldwater fish but have the same disadvantages as the 10-gallon version.

Small 3 gallons and 1-gallon glass bowls are nice as decorations but do not work well for fishkeeping. The only real use for these is to keep bettas in during transportation. There are other small bowls that will work better for short-term transportation of a betta fish, but they should not be considered permanent homes.

Now that we have covered aquarium sizing, let's talk about what types of fish can be kept in which aquarium sizes. A standard 75-gallon tank can be used to keep larger species of tropical and coldwater fishes. It could also be used for a very small reef tank. Keep in mind that the larger the aquarium you have, the more help you will need to keep it clean.

Types of fish

Any of the following fish will thrive in a 75-gallon tank:

Arowana, Cichlids (except angelfish), Large Plecostomus, Large Oscars, Catfish (such as plecos and suckers), large koi and Goldfish. Another benefit to having a 75-gallon aquarium is that you can easily have an assortment of fish at one time. Tanks this size should only be filled about half way for safety reasons; these tanks are very heavy when they are full.

A 55-gallon tank is one of the most popular aquarium sizes. They are usually found in most pet stores and are easily affordable. They can be used for any number of fish, such as:

Fish that would thrive in a 55-gallon tank include the following:

Butterflyfish, Gouramis, Livebearers (such as guppies), Angelfish, Plecostomus, Tangs and many types of large tropical and coldwater fish that love lots of swimming room. There really isn't a downside to setting up an aquarium that could accommodate this list of fish. The main drawback for many is the price of the tank.

Other aquarium sizes that are very popular include: 29-gallon, 20-gallon, 15-gallon, 13-gallon, 10-gallon and 5-gallon. Most of these aquariums will house anywhere from 1 to 4 fish depending on species and size.

If you can't decide which type or sizes of fish to keep in your aquarium, there are some easy ways to help narrow down your options. One way is by choosing a specific type of animal that you would like to keep. This method is great because it takes the guesswork out of choosing compatible animals for any given setup.

Here are some examples:

- Larger Cichlids (such as lab cichlids or angelfish): 75-gallon tank
- Many types of tropical fish (such as gobies, butterflyfish, angels and oscars): 55-gallon tank
- Goldfish and koi: 55-gallon tank or larger
- Tang, Clownfish and Hippo Tang: 29-gallon tank (with lots of live rock) or larger

- Arowana: 75-gallon tank with two water pumps running in opposite directions to create a current

If you have found that you want to keep many different types of fish in your aquarium, it might be worth going with a few different sizes. This way you will be able to keep several different types of fish in each aquarium and have more options for choosing fish later on.

Chapter 6. How to feed different fish

A freshwater aquarium is usually a bit more complicated than an ocean tank, but not always. The best thing you can do is keep the water quality up and feed your fish according to their needs!

Here are some tips for feeding your fish:

Size

The size of your fish will determine the amount of food it will need. Smaller fish (like tetras) will require less food than larger catfish. Keep this in mind if you have a mixed tank with different sizes of fish. Some websites will tell you how much each species eats per day in grams, which is helpful if you're considering what mix of food to get at the store.

Variety

Your fish will get sick if they eat the same thing all the time, which means they may end up dying. Some fish only eat protein, or algae, or veggies. Feed them a variety of food so that they don't get bored and won't develop health issues down the line.

Amount

How much to feed your fish is always a bit tricky. Too much will fill them up and cause waste, too little and you'll have hungry fish that might be tempted to go after each other's fins (or tails!).

Generally, the rule of thumb is to feed 10% of the fish's weight in grams per day. So if your fish weighs 3.5 grams per day, feed it 0.35 grams (0.10 ounce) of food every day.

Frequency

A constant diet is great for newly-purchased fish, but for long-term care you may want to vary their food and how often you feed them. You could feed them only once a day. Or you could give them smaller amounts throughout the day. Just be careful not to overfeed.

Variety in Feeding

You can offer your fish food in a variety of ways. Below is a list of options for you to check out:

Aquarium plants: Some fish love nibbling on aquarium plants, and they provide great cover for the fry! Wild shrimp (also known as Amano shrimp) are especially drawn to plants, so try adding some of these if you've got plant-eating fish like catfish or loaches.

Frozen food: This is a great way to give your fish variety in their diet without sacrificing water quality too much.

Freeze-dried food: Freeze-dried foods are easy to digest, and they're packed with nutrients, so your fish will love you for it!

Live food: While more advanced than dried or frozen foods, live foods have their advantages. Live brine shrimp or worms are loaded with nutrients that your fish can't get enough of. You could feed them live foods every now and then to give them a boost.

When to Stop Feeding

As mentioned earlier, overfeeding is the greatest cause of water pollution in an aquarium. How quickly the food decays will depend on a couple things: what you're feeding and how much of it you give. Most commercially-available fish foods will have a rate of decay listed on the package. If you're feeding both live foods and freeze-dried or freeze-dried foods, then you should check it often to make sure your aquarium isn't getting too cloudy.

Fresh vs. Frozen

Fish can get sick from eating frozen food if they don't have the proper nutrition to process it. Therefore, it's best to feed them only frozen food if they don't have access to live foods at all times (like in a community tank). On the other hand, fish that do have access to live foods will be fine with frozen as well.

Fish Weight

If you notice that your fish are getting too skinny, it's time to step up their feeding. As fish age, they tend to require more food. It's a good idea to feed your fish more often if you know one of them is getting old.

All in all, when it comes to feeding your fish, the key is to give them variety and plenty of options. It may seem tedious at first, but you'll get into a routine after a while!

Chapter 7. How to maintain healthy fish

If you buy a fish tank, it comes with the water and most of the equipment needed to keep your fish happy and healthy. They will need some extra care though, and if you don't provide it they won't survive long! Read on to learn the basics of how to properly maintain a healthy fish.

The first thing you need to do is set up the tank. Put something in the bottom of the tank to prevent your fish from swallowing gravel or other debris. Gravel can hurt their insides and they are not born knowing how to avoid this stuff, so put a couple of flat rocks on the bottom of the tank. You may want to clean them with a mixture of two cups warm water and one tablespoon bleach first though, as bleach is toxic for your fish.

Next, add water and let it sit for 30 minutes before adding any fish (one hour if it's really hot outside). This allows the water to stabilize and come to room temperature. Adding a fish too early can shock it and cause it to die. You want the water in the tank to be between 72 and 82 degrees, so if it's too warm or cold you can add a heater or cooler until it reaches the right temperature.

The next step is adding fish! Be sure you have enough space for all of them. The general rule is one inch of healthy fish needs two gallons of tank space. So, for each inch of fish you have, you'll need four gallons of water. If you can't get a bigger tank, don't add any more fish until you can.

Now that the tank is all set up, you need to decide how often to feed your fish. You should give them small amounts of food several times a day, rather than giving them large amounts of food less often. Make sure to remove any uneaten food within an hour so the water stays clean and free of bad bacteria.

You now have a healthy fish! If your fish becomes sick or injured it is important to treat it properly to give it the best chance at life. If your fish is caught in the tank filter, for example, you'll need to isolate it immediately. Keep feeding like usual and draw water from a different part of the tank so that you don't re-contaminate his water while treating him.

If your fish has an injury or illness, keep it clean and warm. You can use a tea towel soaked in warm water to wrap around its body if needed, but don't use anything else like your hands or clothes as since they could also be sick. The color of the fish's skin will be different depending on how sick it is. In the case of bacterial infections, it can look brownish or greyish.

If your fish is injured and you can't get it to an emergency veterinarian right away, use a mixture of two tablespoons Epsom salt and one gallon water to clean its wound. You can also put a piece of cloth in the tank with the fish on it that's soaked in this mixture to help it keep clean while healing.

Some types of fish are very hardy, including many types that are commonly found in people's homes. Others aren't so lucky and need extra care, but if you take care of them well they should be able to live out their lives with you.

Chapter 8. Diseases and treatments

A fish's health is important for a number of different reasons. For starters, if your fish are unhealthy, they're likely not living their best life. In addition, if the environment in which they live is unhealthy, then it could stress them out. This might make them susceptible to more illnesses. This is important, because fish who are constantly sick tend to die more often. So, keeping fish healthy is a lot like keeping fish alive. Its also important for the hobbyist, because an unhealthy tank won't look as good or last as long as a healthy one.

The first step in keeping your fish healthy is understanding what is causing problems in the first place. This involves knowing which illnesses commonly afflict your variety of fish. It also requires that you have some idea of why they get sick (i.e., how the illness takes place). From there, you can take various steps to protect them and help them recover when they're ill (if possible).

The diseases and symptoms listed here are the ones that were seen in the fishkeeping press. They are not necessarily the ones that anyone else has seen. For this reason, always make sure to consult your local fish veterinarian, as many diseases can be deadly if left untreated.

Fish Diseases: Causes and Diagnosis

Diseases will take hold of fish in different ways. Not all will spread through contact (as is commonly thought). And not all are transmittable between species (i.e., they vary from species to species). So, if one of your fish dies, then it's important you start checking the water immediately to see if others of your kind have been infected.

Sea lice will take hold of fish in different ways. These are parasitic creatures that resemble tiny shrimp. They live by attaching themselves to a fish's skin and feeding on it (hence the name "parasite"). Sea lice typically don't do any harm to the fish, but they certainly do cause problems for you. Their presence can be hard to spot, and they're also difficult to get rid of. However, because they tend to stick around for quite some time before disappearing, it's crucial that you find them before too many other issues arise (i.e., fish dying).

There are a number of different things that can cause your fish to lose their color (i.e., turn white, gray, or black). Some of these things include:

Anemia: Probably the most common cause of color loss, anemia is usually the result of parasites attacking the fish in some way. The reason this causes skin discoloration is because when blood cells are damaged or destroyed, it will leak into the surrounding tissue. Because it has iron in it, blood will damage the tissue around it and change its appearance from pink (when healthy) to gray or black (when damaged).

Probably the most common cause of color loss, anemia is usually the result of parasites attacking the fish in some way. The reason this causes skin discoloration is because when blood cells are damaged or destroyed, it will leak into the surrounding tissue. Because it has iron in it, blood will damage the tissue around it and change its appearance from pink (when healthy) to gray or black (when damaged). Stress: A number of different things can cause stress in fish (i.e., overcrowding, poor water quality). And one of these things is usually parasites. So, if your fish are stressed for some reason, they're more susceptible to illnesses and parasites. One such illness that often results from stress is anemia. This can cause the same sort of discoloring as above.

A number of different things can cause stress in fish (i.e., overcrowding, poor water quality). And one of these things is usually parasites. So, if your fish are stressed for some reason, they're more susceptible to illnesses and parasites. One such illness that often results from stress is anemia. This can cause the same sort of discoloring as above. Starvation: In a number of cases, the old saying "starve a cold" (meaning starve a cold to death) is true. This will cause anemia in the fish because when iron becomes scarce in their bodies, the food they eat will also become scarce (i.e., "starve" them). This can be the result of poor water quality or just having too many fish. Either way, if your fish are starving, they will likely soon turn black.

In a number of cases, the old saying "starve a cold" (meaning starve a cold to death) is true. This will cause anemia in the fish because when iron becomes scarce in their bodies, the food they eat will also become scarce (i.e., "starve" them). This can be the result of poor water quality or just having too many fish. Either way, if your fish are starving, they will likely soon turn black. Iron Deficiency: Not all parasites need to make their hosts sick first before feeding on them. Some will do it right away. Fish, like people, can suffer from iron deficiency (i.e., anemia). Because their bodies are lacking in iron, this will cause red blood cell damage and death. This is one of the most common causes of anemia in fish and is generally a result of overstocking or poor water quality (polluted water).

Not all parasites need to make their hosts sick first before feeding on them. Some will do it right away. Fish, like people, can suffer from iron deficiency (i.e., anemia). Because their bodies are lacking in iron, this will cause red blood cell damage and death.

Chapter 9. Water chemicals and plants

Different types of water

The type of aquarium water you have can make or break the health of your fish. The two most common types are freshwater and saltwater. Freshwater is the type of water that falls from the sky, while saltwater is found in oceans and other bodies of water.

Differences between fresh and saltwater

Freshwater aquariums can be maintained with chemicals such as chlorine or chloramine, but saltwater requires a different set of chemicals including a salt. Chemicals are necessary to control certain levels, such as pH balance, which ensures that your tank doesn't get too acidic or basic for your fish to live comfortably in it. The other difference is that saltwater aquariums need to be refilled with water every week or so, while freshwater aquariums need to be refilled only when damaged.

Your fish do not like the salt in the water, and can easily die because of it. Also, your tank will get a bad smell from the salt. When you add saltwater to fresh water, it will evaporate immediately and could cause some damage to your tank. While salt is dissolved in fresh water to maintain proper pH levels, some animals require a certain amount of magnesium chloride in their diet. If you add too much quantity of salty water into an aquarium you might also cause harm due to excess minerals entering your fish's systems and causing irregular heartbeats.

Water

Water is an important part of any aquarium, but it doesn't have to be exactly like the water that your fish came to you in. As an aquarist, you will need to change some of the water from time to time, and you may need a certain type of water depending on the fish in your tank. Freshwater can be replaced with distilled water or rainwater. This will help reduce stress on the fish as he adjusts from his previous environment and replace minerals that were depleted while cycling the tank. Saltwater is generally replaced with sea/ocean water for health reasons as well as conservation options.

Other Chemicals

Algaecides - Algaecides are used to kill algae in order to keep the tank water clear and provide optimal living conditions for the fish. You should make sure that you do not overuse this chemical as it can also kill fish, so you should know how much your tank can handle.

Aquarium salt – Aquarium salt is used mainly for changing the ph levels of some fresh water tanks. It can also help prevent certain diseases or help with rehabilitation from disease. You should research your fish to see if it requires this chemical before adding it to your tank.

Peat - Peat is another chemical used in aquariums that helps keep the water clean and prevents certain unwanted substances from entering into the water. You can get peat from a garden supply store.

Water Changes – Water changes are done to replace the water that evaporates or is taken out of the tank. It might be done every few days or more often depending on how much water your fish use and how big of a tank you have.

Water Conditioner – Water conditioner is used to help remove excess chemicals, dirt and other particles from your aquarium. You should know the type of conditioner you will need for your tank before making any purchases as some might be harmful to your fish. Also, if you have gravel, you should add some tap water into it as this will also help clean the gravel.

Water Treatment – Water treatment is used in aquariums to treat the water so that it can be used for fish. It can be bought online depending on your tank size and type.

How to chose plants

Choosing the right plants for a freshwater tank may seem complicated at first, but don't worry, it's not as difficult as you might think. The basic rule is to always check what your fish species prefer to eat. In general, carnivorous fish like Siamese algae eels or large Pleco-like catfish are better suited for tanks with fewer plants. On the other hand, goldfish and koi will usually prefer an abundance of algae so they can graze on it in between feeding time.

In general, beginners should choose aquatic vegetation that is easy to maintain, like Java moss (Vesicularia dubyana) or hygrophila. You can also use artificial plants and decorations, like those made out of silk. In this case, you will have to clean them from algae regularly or replace them with new ones... But most fish species will only eat the leaves anyway!

Set up your aquarium plants

Before you add the first plant to your aquarium, make sure that all of the equipment is working properly. Check for leaks and cracks in the tank, filter and aerator. Fill up the tank with water and let it run overnight to check for leaks.

Use water conditioners

If you buy live aquatic plants, make sure to use water conditioners to avoid yellowing leaves. Always check what your fish species prefer to eat. In general, carnivorous fish like Siamese algae eels or large Pleco-like catfish are better suited for tanks with fewer plants. On the other hand, goldfish and koi will usually prefer an abundance of algae so they can graze on it in between feeding time.

In general, beginners should choose aquatic vegetation that is easy to maintain, like Java moss (Vesicularia dubyana) or hygrophila. You can also use artificial plants and decorations, like those made out of silk. In this case, you will have to clean them from algae regularly or replace them with new ones... But most fish species will only eat the leaves anyway!

Chapter 10. The Nitrogen Cycle of Your Aquarium

The nitrogen cycle is the main cycle that you must bother about initially. Many hobbyists and beginners don't bother about the nitrogen cycle and they have to face many dilemmas in the future in terms of maintenance and aquatic diseases. The nitrogen cycle in your aquarium is a crucial process to maintain the amount of beneficial bacteria in your aquarium and in the filter media that will help in the conversion of ammonia to nitrate and then conversion of nitrate to nitrites.

Organic waste products in your aquarium result in nitrogen pollution of your water. Some common waste products in your shrimp aquarium are excretes of fish, uneaten food products, rotting algae, and dead creatures. Nitrogen resides in the aquarium in different forms. Nitrogen containing waste products in the aquarium are available in the form of waste proteins, ammonia, ammonium, nitrate, and nitrites. The waste products inside your aquarium contain proteins that convert ammonium into ammonia with the help of biological decomposition. Now you can measure the ratio between relatively harmless ammonium and toxic ammonia with the help of pH value. The higher the pH value, the more ammonia is formed inside your aquarium.

You must test the levels of ammonia, nitrate, and nitrite before putting your tropical shrimps inside the tank. You can test your aquarium water every alternate day if you have a small aquarium or every week if you have a large aquarium. Note down the reading of ammonia, nitrate, and nitrite levels. Initially, the levels of ammonia would rise. After few weeks, you should monitor the rise in nitrite levels and a slight decrease in ammonia levels. Finally, after few more weeks, the level of nitrite will start declining and nitrate levels will gradually rise in this period. When you no longer detect ammonia or nitrites but only nitrates, then this condition is assumed safe for adding tropical shrimps to your aquarium. The best way to introduce your shrimps to your aquarium is to introduce them one by one. Initially, you can introduce some native shrimps that can bear changes in chemical composition and keep reading the levels of ammonia, nitrite, and nitrate in your tank. Then, introduce other tropical shrimps one by one by balancing the levels.

Different Stages in Nitrogen Cycle

Nitrogen cycle involves many stages that you must understand to maintain the ecosystem inside your aquarium. Following are the stages of the nitrogen cycle: --

- Stage1: -- Ammonia is introduced into your aquarium by tropical shrimp waste and uneaten waste food. Now, the shrimps waste and excess food would either produce ionized ammonium or un-ionized ammonia. Ammonium is not harmful for tropical shrimps but ammonia is and you have to control its amount. Now, it depends on the pH level of the aquarium water, whether the waste should change into ammonium or ammonia. If the pH level of your shrimp tank is 7 or higher, then the wastes should convert into ammonia. But when the pH is lower than 7, the wastes should convert into ammonium. This is the main reason why you must measure the pH level of your aquarium water often.

- Stage2: -- In this stage, some beneficial bacteria called nitrosomonas will develop and oxidise the ammonia presenting in your tank; thereby making nitrites as a by-product. Although, we no longer have harmful ammonia inside our aquarium, but still nitrites are also not beneficial for your shrimps. Nitrites are as harmful as ammonia to our tropical shrimps.
- Stage3: -- In this stage of the nitrogen cycle, bacteria called nitrobacter converts nitrites into nitrates. Nitrates are not as harmful as compared to ammonia and nitrite, but still a large amount of nitrates is not permissible for your tropical shrimps. The quickest way to get rid of excess amount of nitrates from your aquarium is a partial change of water. Monitor the nitrate

levels and do a partial water change when the nitrate level exceeds permissible limit.

Ammonia and nitrite is not good for your shrimps and you must control them to minimum limit. Anything below 5mg/l is permissible for your shrimp aquarium. If you monitor a gradual increase in any of them, then reduce the amount of shrimps inside your tank, check or clean your filter, and you can also use sera ammovec that is a chemical you can buy from your nearest aquarist marketplace. The permissible limit for nitrate is not more than 20mg/l and anything above it should be avoided. Immediately change a partial amount of water from your shrimp tank if the amount of nitrate increases above the permissible limit or reduce the number of tropical shrimps from your aquarium.

Chapter 12. Taking care of fish

Diet & Nutrition

Like humans, most beginner freshwater aquarium fish need a variety in their diet to stay at their best. Different fish need different kinds of nutrition, so which should you use for your new menagerie? The vast majority of freshwater aquarium fish are omnivorous, which means they need a diet of both protein and plant matter. The considerable amount of options in the fish food aisle can be overwhelming, and there are many choices when it comes to diets for your fish friends, so here I will break down different types of food.

Flakes

Fish flakes are definitely the most common type of food in the aquarium trade. It is an everyday staple in the diet of many aquarium inhabitants. While this is often the most cost-effective option, you should not merely look for the cheapest flakes because they differ in quality. Make sure you check out the label for the list of ingredients. The list is organized by the amount of content for each item. For example, if the first ingredient is 'shrimp,' that means that the flakes are highest in a concentration of shrimp. This is good because you know that the primary ingredient is comprised of protein. If the first ingredient is anything other than something that contains protein – stay away! But well-balanced flakes are key. The ingredient list should start with two to three protein ingredients, followed by a few plant-based ingredients. Some unpronounceable ingredients should be expected, as they are generally preservatives. Still, too many unrecognizable things should be a red flag.

Freeze-Dried Foods

Freeze-dried foods are my most preferred method of supplementing protein. As you will read below, live foods can contain pathogens and diseases, and frozen foods can cause fluctuations in temperature if not prepared properly. These foods are also great because they can be broken down into tiny pieces to fit the small mouths of many beginner freshwater fish. Both flakes and freeze-dried foods can be ground down into tiny pieces. This means that even the most miniature nano fish can get the proper diet that will make them thrive. Additionally, both flaked and freeze-dried foods will keep fresh for many months, so you can save money by buying in bulk!

Frozen Food

Many freshwater fish will snub flaked food or only eat it if they are starving. This is uncommon in most beginner fish, but it depends on the species and how they were raised. Frozen food is a fantastic option for these finicky eaters! Live foods come with risks, especially for a first-time fish keeper, so frozen foods might just be the key to your fishy's health. For those species that need a ton of protein to thrive, frozen food is optimal. It will stay fresh for several months, provides protein that your fish need, and carries little risk of having diseases, pathogens, or pests that can infect your tank. Just portion off the amount your aquarium can eat in a few days in your freezer to let it thaw, then leave it out for an hour or two to reach room temperature before adding it to your tank. Your fish are sure to devour this treat, and with proper preparation, it carries little risk.

Live Food

Some freshwater fish only thrive with live foods. But, as mentioned above, live foods do carry some risks. While it is rare, live foods such as daphnia, bloodworms, and brine shrimp can carry pathogens and diseases that can easily take root in your tank. You can do little to prevent this other than by making sure to buy your live foods from established businesses with respectable reputations. It is the best way to try to ensure disease free food.

Quantity and Schedule

It can be challenging to know how much to feed your new fish friends. They will typically continue eating for as long as they can. Some species will even eat so much that their stomachs rupture, which often causes death! Take care to only feed your fish as much as they can eat in about five minutes.

Fish are usually purchased as juveniles, and young fish need to be fed more often than when they are full-grown to ensure they reach their maximum size and best coloration. You should start out feeding them small amounts three to four times per day – no more than they can eat in around three minutes. After a few months, you can reduce the number of feedings by one. Once they have reached their full size, you can feed them one to two times per day, and as much as they can eat in five to six minutes.

Be sure to remove any uneaten food from the tank once the allotted time is up. This will prevent your fish from overeating and help keep your tank clean. Uneaten food can sink to the bottom of your tank and become trapped around plants and decorations and sink down into the nooks and crannies of substrate. This will pollute your water and creates a breeding ground for bacteria. So, do your fish friends a favor and make less work for yourself down the line by removing uneaten food!

Common Beginner Care Mistakes

Many pitfalls can trip up new aquarium owners. Listed below are some common mistakes and myths that inexperienced aquarists fall victim to. The greatest assurance against failure is education – so be sure to do your research!

The Fish Will Grow to the Tank Size

Many people have the mistaken idea, for some reason, that a fish kept in a small space will stop growing before it gets too big for its tank. I am unsure of where this mistaken idea comes from, but it is patently false. All fish species have a general adult size they will grow to, and it is essential to know how big your pets will get when fully grown to ensure you give them a proper tank size.

Over or Underfeeding

As stated above, overfeeding is an easy mistake for beginners to make. It can cause your fish to become ill or even die. But underfeeding can be just as big of a problem. Without enough nutrients in the proper proportions, your fish can also become sick and die. Even if they do not die, they will be lethargic and dull in color if they are starving.

Improper Medications

Different ailments of your fish will require different medications. It is of the utmost importance that you read the instruction label carefully and follow it exactly. Improper dosage can actually do more harm than good. If you have doubts about the kind of medication you need or the illness your fish has, be sure to consult with a veterinarian experienced with freshwater aquarium fish.

Not Cycling the Tank Properly

These next two mistakes are often the most commonly made among beginner aquarists. To make sure your tank has a healthy nitrogen cycle established is imperative to your pets' health. It takes some time for the tank to cycle, and many beginners are impatient and add fish too early. Doing so puts their lives in jeopardy.

Overstocking

Having a new tank is super exciting! You have put in so much hard work, energy, effort, and money to get everything ready, waited patiently while the tank cycled, and made absolutely sure everything was perfect for a safe and comfortable home. Now, when it is time for you to fill your tank with beautiful new life, it can be tempting to want to share your home with as many as possible. However, adding too many fish to your tank will put stress on them and create more work for you. Overstocked tanks are not a healthy environment for your fish and will substantially increase the amount of maintenance you must do to keep it clean.

Using Small Tanks or Bowls

I would never recommend using anything less than a 3-gallon tank, and that small only in cases of a single betta fish, although a 5-gallon would be preferable in that instance. Freshwater aquarium fish need space to swim and explore. Many are schooling or shoaling fish that need to be kept in medium to large-sized groups of their own kind. A tiny tank, or even worse, a bowl, is not nearly enough space to make a comfortable home for any fish. Any aquarium inhabitant will surely die prematurely if kept in such conditions.

Omitting Equipment

While some aquarium equipment may be substituted for other alternatives, omitting essential equipment such as a filter, water pump, and comfortable décor is irresponsible. Without the needed hardware, your tank will quickly become fouled and dirty. Without comfortable décor, your fish will become stressed. Without both, they will likely die.

The only way to have happy fish is to have healthy fish. Taking care of them properly will be so much more rewarding for you, as they will be active, fun, and full of color to brighten your home. Every effort you put into giving them the best home possible will be paid back tenfold. Now that you know about feeding, maintenance, and common mistakes for beginners, you have all the tools at your disposal to become a master aquarist and contribute to the fantastic community of aquarists from around the world.

Chapter 13. Technology and Equipment for Freshwater Aquariums

Basic Technology and Equipment

Most aquarium owners will tell you that their method for caring for fish is the best. If they didn't feel this way, then they would probably change their method.

One important prerequisite to remember is your aquarium budget. There are quite a few effective and fancy ways to run an aquarium, but they may be difficult and expensive to start and require large amounts of space.

Quite a few aquarium shops will sell starter packages that include all of the fish you need to own. Unfortunately, the equipment in the starter packages may be poor quality, cheap, and even obsolete. Remember, you will get what you pay for - within reason.

Most of the air pumps that you will find in these packages would wake a graveyard. The gravel is often dyed and will leach into the water. The starter kits may also include cheap plastic decorations that ruin the natural appeal of an aquarium environment.

Make sure that any frame you buy is stable enough to hold the weight of your tank. Make certain that the glass is strong enough to hold the weight of the water in a larger tank. If an aquarium is thin, cheap, and big, then you need to shop somewhere else.

Koi

Aquarium Stands

Your aquarium will need to go on a stand. Since water is heavy, you may want to buy a stand specially made for your tank. If you have a small aquarium, you can place it on existing furniture in your home.

Remember that a gallon of water weighs 8.34 pounds, or 1 L weighs 1 kg.

This is just as relevant when placing a tank upstairs, in an apartment, or on floorboards.

You don't want an aquarium coming through the ceiling in your house. Place a heavy tank so that its weight is supported evenly on several joists at 90° to the tank rather than just a single joist. It is even better to place your aquarium against a load bearing wall.

Take care to make the stand level. Wood can often warp, and not every stand is built perfectly. To keep your tank level, you can place a layer of polystyrene foam between the tank and the stand. This will help the weight to distribute more evenly and could prevent breakage of the tank,

Temperature Controls

Since fish are cold-blooded, they are highly sensitive to heat and cold. You need to take every precaution to keep a stable temperature in your aquarium for this reason.

Heaters

Hydor ETH In-line Heater

A brackish aquarium system will require a heater. A heater can keep your tank at a stable temperature.

There are two basic types of aquarium heaters:

Submersible Heater: This heater stays completely below the water and is the most common type of heater in most aquariums. The heater is made of a glass tube surrounding a metal coil, which will provide heat.

Most submersible heaters come with a thermostat that allows you to turn the heater on when the water gets cold and off when it reaches the correct temperature.

Azoo Titanium Heater

These thermostats operate with a metallic switch to regulate the current. The heater will have a dial on the top of the glass to show you the temperature. Other heaters may have a dial on the power cord outside of the water.

In-Line Heaters: These heaters heat the water through a pump. They often operate with an external thermostat, which runs with an integrated circuit set up outside of the tank, similar to a small computer.

These heaters are often used in larger aquariums since they can heat large volumes of water.

3 W per gallon, or 0.75 to 1 W per litre, should be ideal for an unheated room. Use 2 W per gallon, or ½ W per litre, in a room that is unheated. It is important that the temperature of the tank is consistent day and night.

Submersible heaters are best used with two smaller heaters than just a single large heater. If you require 300 W for heating, it is best to use two 150 W heaters. If one of the heaters stops working for any reason, the other heater can support the temperature of the tank. Likewise, if one heater does not turn off at the right time, the other will so that the tank does not overheat.

Heaters will fail.

Most submersible heaters are made with cheap parts and unsophisticated technology. When the thermostat stops working correctly, you will have a very hot or very cold tank because the heater won't shut on or off when it is supposed to. Even with more expensive submersible heaters, you can have issues with the thermostat switch.

To use all heaters safely, make sure that they comply with your national or state safety regulations. You do not want your fish, let alone yourself, to get electrocuted due to a cheap heater. Cheap heaters often have larger currents, although it is possible to get a heater with low voltage at 24 V as opposed to 110 or 240 V. However, these are more expensive heaters.

Chillers

Just like they sound, chillers will be used to chill the water to bring it down to the right temperature. Chillers are ideal for cold water tanks year-round and some tropical freshwater tanks in warmer weather. Fish do not normally do well in water at high temperatures.

Since an aquarium is a closed system that has a limited water volume, the water has the capacity to heat up more quickly than in the wild. This is very true in warmer areas. If your aquarium is not in an air-conditioned room, the fish could overheat.

Chillers are also used to cool deep water tanks and create a very cold habitat for fish that live at great depths.

A chiller will work with the same technology as a refrigerator. It will use the expansion and compression of gas to cool an aquarium when water moves through the chiller. A chiller is controlled through an external thermostat that will regulate temperature.

Oceanic Aquarium Chiller

Thermometers

A thermometer is critical so that you can monitor your water temperature. There are several types of thermometers available for aquarium use. The first type is a standard glass bulb thermometer that contains a column of alcohol.

This thermometer may come with a silicone or rubber suction cup to stick it to the inside of an aquarium. This is a good thermometer in that it is user-friendly and inexpensive.

However, this thermometer is not always the most accurate compared to a more expensive thermometer that is calibrated to have a higher standard of accuracy.

Mannix AQ150 Digital Aquarium Thermometer

Nonetheless, these thermometers will tell you the temperature of an aquarium within at least a degree. So if you stay within an acceptable range on a consistent basis, your water temperature should be fine. Some glass thermometers may also have an integrated hydrometer, which I will discuss later in the book.

Thermometers with a plastic adhesive strip that will display the tank temperature are also available. These thermometers provide a good indication if something is wrong in the water temperature of the tank.

These thermometers are also user-friendly and cheap but can be imprecise.

The most accurate thermometers for a freshwater aquarium are battery operated or electrical. These thermometers will have LCD displays and use probe sensors to read the temperature of the water. A high-quality thermometer will also provide you with an outside ambient temperature. This outside temperature will help you to determine how much you need to heat or cool your tank in the event of a climate change.

Now that you understand the basics of the technology and equipment that you need to successfully run your aquarium, it's time to move onto the next step... That's right - filtration.

Understanding filtration for freshwater aquariums will make it easy for you to choose the most efficient filtration system to fit within your budget. Proper filtration will keep the water in your aquarium clean and safe so that your fish remain protected at all times. A clean fish is a happy fish...

Angelfish

Chapter 14. The Correct Stand For Your Aquarium

A stand is another vital consideration you cannot overlook.

Your aquarium needs to sit on a stand that's strong enough to support its weight. After all, you don't want to come home one day only to find your aquarium destroyed because your stand gave way.

When searching for the most appropriate stand, account for the specifications of your aquarium.

What to Think About When Considering Aquarium Stands

Because you cannot just place your aquarium anywhere, here're some of the most important elements you need to consider as you choose one:

The size, weight, and shape of the aquarium

The size of your water tank will determine the size of your stand. It's evident that the bigger the tank, the bigger a stand you'll need. However, even though the size of your aquarium is the first aspect to consider, you should not overlook the weight.

Water is heavy, which means you should get a stand that can support the weight of your aquarium once it's full of water.

Don't just grab any furniture you have in your home and use it as a stand. If it's not strong enough, it will cave under the weight of your water tank. The strength of the stand should match the gallon capacity of your tank.

The shape of your tank is another point to consider when determining which stand to buy or assemble.

Can a triangle stand be effective for a rectangle water tank? NO! Some stands suit rectangular tanks, while others work best with aquariums that have bow-shaped fronts.

Because of this, you must identify a tank suited to the shape of your aquarium.

Understand the different materials used for a stand

There are different types of materials you could consider for your stand. The most popular materials used are:

1: Particle board (MDF)

MDF is an inexpensive type of material normally used when strength and appearance do not matter as much as cost.

Made from mixing sawdust with adhesives to hold together, although the particle board is light and weak, it comes in different densities that can withstand different strengths and resistance.

A major disadvantage of this board is its tendency to expand and discolor after absorbing moisture, especially when not covered with paint or a suitable sealer.

It is, therefore, rarely used in places that have high levels of moisture, making it unsuitable for aquarium stands because water will occasionally spill onto the board during aquarium maintenance. Spillage will then cause the board to soak in water, which will cause it to expand, weaken over time, and begin to crumble.

2: Plywood

Plywood is a material made up of thin layers of wood veneer glued together. A veneer is simply a thin slice of wood that is thinner than 3mm. The strength of plywood increases after the gluing together different veneers to produce flat panels such as cabinet tops and parts of furniture, including doors.

Plywood is a better choice as it is more durable and strength resistant than the particle board. It also much stronger and can hold heavier tanks than the board can. The glue used to hold together the veneers that make up the plywood is waterproof, and hence, it's less susceptible to water damage, which means it can last longer.

Damage, however, can happen to the plywood if exposed to water many times. However, with this material, if water soaks into the wood, you can dry it by encouraging evaporation without much damage to it.

3: Metal

Metal stands are the best for large aquariums since they can handle a great deal of weight. Metal stands are durable and not easy to damage. A more open metal stand will give you the advantage of increased access to your back end electronics.

The most common materials used to make a metal stand frame are stainless steel or aluminum. Stainless steel does not rust, making it an excellent choice for a stand.

Aluminum is more popular outside the US but is gaining recognition because of being lightweight. Aluminum is ideal also because you can easily custom-make a unique stand at a lower price—lower compared to building one from wood.

Metals rust and there are many ways of preventing it. Many aquarium owners prevent rust by powder coating the stand in nylon, vinyl, epoxy, polyester, acrylic, or urethane.

To do this, first ensure that your stand is clean and dry. Then, spray the dry powder and then heat it, thereby turning the powder into a thin film. Painting the stand is another standard option.

Please note that no finish or paint is 100% rust proof, which demands that you maintain your stand very well by ensuring it's always clean and dry to prevent rust from forming.

Space for equipment

You'll require enough space in your stand to store all the equipment you need for aquarium maintenance. A reef aquarium, for example, requires a lot of equipment and, subsequently, needs a stand that's large enough to store the various tools.

To keep your aquarium beautiful and in tip-top shape, it is best to get a stand that has closed a space that you can use to hide the equipment. If you'll want to add a sump into your aquarium, then a stand with concealed sides, with holes at the back of the board through which wires pass through, revealing a tidy aquatic space, would be an excellent option.

Opening and height of your stand

Generally, a stand remains open at the top, which works well for glass aquariums. If the glass is rimless, the top of the stand will need a self-leveling mat to hold the entire bottom of the water tank because doing otherwise might cause undue pressure to the glass, especially when you're using a wooden stand.

For acrylic stands, the entire bottom of the water tank will need support. An acrylic stand can have an open-top, especially if you intent to install a sump inside it.

You also need to ask yourself if you will be doing much of your viewing from a sited or standing position. Doing this will help you determine the height of your stand so that you can enjoy watching your fish. Most stands have a sited viewing angle whose height is no more than 30 inches. Taller stands are usually between 30 to 36 inches.

The desired, overall look of your aquarium

There's a particular desired overall look you wish to achieve. How your stand looks will either compliment your aquarium or contrast the overall look. Do you prefer a sleek, modern, aesthetics, rustic wooden or sport vintage?

Considering these aspects will help you choose an ideal stand:

Chapter 15. Safety

While many of the following safety tools and suggestions are just that, suggestions, the two you must always try to remember are drip loops and a level floor. The others, while they can be useful in keeping you safe, are not as widely known or used, even among veteran aquarium keepers.

Drip Loops

- With an aquarium, especially with multiple power cords hanging down the back, safety is an important issue. Say you lift the heater out of the tank during your weekly maintenance, the idea being not to have an electrical item in the water when you put your hand in the tank, and some water runs down the power cord to the power outlet or power strip/ extension cord etc. This could cause a shortage etc. and damage your electrical outlet. Drip loops prevent this. A drip loop is essentially a loop in the power cord just before the cord plugs into whatever electric outlet or power bar is being used. A power bar would have to be placed above the floor in some cases to prevent the water running straight down the wire to it. This loop prevents water from going straight down the power cord and into the outlet. Take a look at drip loops online to get an idea of what they look like.

- Another possible safety measure is the purchase and use of a grounding probe. See if you can also cover your electrical outlets and power bars with something to prevent water from landing on them. Or simply make sure your electrical outlets and power bars are GFCI and also possibly surge protected. GFCI outlets and power bars are specifically meant to protect against wet and humid conditions in which users can otherwise get shocked.

Surge Protectors

- To protect your equipment against electrical surges you should ideally purchase and use a power bar that has a surge protector feature instead of using an ordinary power bar or directly plugging equipment to your wall power outlet. This is however not a necessity and many hobbyists do not bother to use them either because they do not know about them or because they cost more than normal power bars.

GFCI's

- While many if not most people in the hobby do not know or do not bother with them, GFCI's are devices that protect people from electric shocks when water is involved. Simply put, this device stops the current from flowing through the outlet

(whether it's a wall outlet or a power strip/ power bar that has a built in GFCI) to your wet fingertips and shocking you.
- If you have the skill, there are GFCI outlets that you can purchase and install, replacing your normal wall power outlet, to protect yourself against electric shocks.
- You could also purchase a power bar or power strip or extension cord that has either a GFCI or some other protective feature.

Grounding Probes

- These probes are mounted on the aquarium and plugged into your wall GFCI power outlet. The idea is that if for example a heater in the aquarium becomes old or damaged and leaks a current or a light on top of your tank falls into the water because someone bumped it while you were away, the current would go through the probe to your power outlet, tripping the GFCI.
- The alternative would be you putting your hand in the tank and getting shocked yourself if a grounding probe was not being used.
- Many aquarium keepers do not know of or choose not to use grounding probes. One of the reasons being they would like to know there is a problem instead of the GFCI taking care of it without their knowledge. In this case, they willingly risk

getting shocked. While many items in the aquarium hobby would not seriously harm a person with the level of current they would emit, they still pack a punch. Another reason is that they do not want to create a closed circuit in the tank and possibly harm the fish. Either way, the human then takes on the risk.

- Again, many people do not use grounding probes, either due to not knowing about them or choosing not to use them, and they can often not experience anything negative by doing so. If you keep your equipment well maintained and ensure they are quality products with little to no damage over time, while you do not eliminate the danger, you do reduce your chances of being exposed to it.

Level Floor

- Make sure that your floor is level and not warped or slanted. Buy a level (like a beam level or laser level etc.) and make sure your floor is within the acceptable range. A level floor will prevent a lot of headaches, especially with larger tanks. The larger the tank, and more uneven the floor is, the bigger the risk of the tank's seals coming apart or the tank glass simply cracking and all your water and fish ending up on your floor.
- If you place your aquarium stand and aquarium on an uneven floor, you are putting pressure on the joins of the stand and the

seals on the aquarium depending on which way the floor is tilting or sloping. This can cause seals to weaken over time and glass to crack once the strain on the glass becomes too much. If for example you floor is sloping down to the right, that means there is more water moving to and settling on the right side of the tank, which in turn puts more pressure on the right side of the stand to keep up that extra water compared to the left.

- If your tank springs a leak one day or outright cracks and empties the contents of the tank onto the floor, you will have lost all your fish in addition to having to repair your floors and any furniture in the vicinity, repair the walls etc.
- Many aquarium keepers use wooden or composite shims to level their stands when the floor is uneven by placing them below the legs of the stand or anywhere along the length or width of the tank as needed to prop up the stand and make it level before placing and filling the tank.

Many people choose to ignore this aspect of their aquariums but remember, while small tanks may last longer or even arguably never crack from being placed on an uneven floor due to less water pressure, the larger the tank, the bigger the risk of tank seal failure over time and the bigger the headache when it does.

Chapter 16. Problem Solver Chart

The following shows the common problems that occur in aquariums with possible causes and solutions.

Stressed fish after water change

Gasping for air at surface, slows down, stays at bottom and sudden death of a fish.

- Cause/Solution 1 - Make sure water is treated properly for chlorine or chloramine. Check dosage of water treatment conditioners and if it is still effective. Your water supplier might have changed from chlorine to chloramine.

- Cause/Solution 2 - Overdose of water treatment conditioners can take up oxygen. Check with manufacturer's product information on treatment levels.

Gasping for air at surface by fish

- Cause/Solution 1 - Water. Can be caused by high ammonia, nitrite, and nitrates. Check with water tests to confirm. Correct with water changes and/or correct biological filtration.
- Cause/Solution 2 - Gill Flukes if water conditions are good. Treat with medication for gill flukes. Treat aquarium and isolate affected fish if possible.

Slow fish activity

Stays on the bottom or has low appetite.

- Cause/Solution 1 - Can be caused by high ammonia, nitrite, and nitrates. Check with water tests to confirm. Correct with water changes and/or correct biological filtration.
- Cause/Solution 2 - If water conditions are good, possible internal parasites such as worms. Severe symptoms are sunken body. Treat with medication for internal parasite that must be eaten to be effective.

Fins rotting and blood on edges

- Cause/Solution 1 - Can be caused by high ammonia, nitrite, and nitrates. Check with water tests to confirm. Correct with water changes and/or correct biological filtration. Fish may be able to heal itself after water correction.
- Cause/Solution 2 - Bacterial infection due to stress from poor water conditions. Correct the water conditions and treat with anti-bacterial medication. Beware this will kill good bacteria and need to be replenished after treatment is completed.

White dots (ich) or velvet coatings that are on the fish's body and fins

This symptom is known as ich or velvet disease infection.

- Cause/Solution 1 - Treat with medication for external parasites. Isolate affected fish if possible in a separate container and treat. Also treat the unaffected fish to eliminate the parasites in the aquarium.

Tails clamp

- Cause/Solution 1 - Water. Can be caused by high ammonia, nitrite, and nitrates. Check with water tests to confirm. Correct with water changes and/or correct biological filtration.

- Cause/Solution 2 - Velvet. Looks like a hazy milky coating on tail. Treat with medication for external parasites. Treat aquarium and isolate affected fish if possible.

Belly becomes flat or sunken

- Cause/Solution 1 - Parasites, internal worms. Treat aquarium and isolate affected fish if possible. The treatment needs to be fish food treated with medication to be effective. Feed medicated food to all fish. Note that medication kills invertebrates such as snails and shrimps.

Fish head points up, spins and uncontrolled swimming ability

- Cause/Solution 1 - Internal parasite or disease that has affected the brain or nervous system. This is untreatable and the fish should be removed.

Snails migrate and stays on top or pull into their shells for long periods

- Cause/Solution 1 - This is a sign of poor water conditions. Check with water tests to confirm. Correct with water changes and/or check biological filtration.

Snail shell becomes thin and brittle

- Cause/Solution 1 - Usually an indicator of high ammonia levels causing a high acidic environment that eats away the calcium in the shells. Water changes are needed. Also check if your tap water contains chloramines and the water conditioner treatment neutralizes chloramines and not just chlorine. Some water treatments can deteriorate with time and needs to be replaced.
- Cause/Solution 2 - The high ammonia levels can also be due to the filter not having beneficial bacteria to breakdown ammonia. Using beneficial bacteria treatments can be used to correct this. Also check if your filter has good water flow and is working properly. Use zeolite chips in a box filter for a week to absorb ammonia until the biologicals in the filter can catch up.

Cloudy water or green water

- Cause/Solution 1 - Algae bloom due to high ammonia and nitrate levels. Try using the different methods mentioned in this book to reduce ammonia that eventually become nitrates that feed algae. Weekly water changes will also help prevent this by lowering ammonia, nitrite and nitrate levels.

Algae growth is excessive

- Cause/Solution 1 - Snails and algae eating fish can be a possible solution. Mechanical ways such as scrapers can be used, but this will cause additional debris so a filter pad cleaning and water change may be needed after the algae are scrapped off.
- Cause/Solution 2 - Common causes are due to excess nutrients such as ammonia and nitrates. Intensity and length of light also contributes. Reduction of these factors can help but may also affect the health of live plants.

Filter not processing ammonia

Filter not processing ammonia indicated by testing for ammonia levels every few days. Ammonia level needs to be below 0.02 PPM (mg/L) for a healthy aquarium.

- Cause/Solution 1 - Often indicated by continued high ammonia levels above 0.02 PPM. If changing water and filter material doesn't help the high ammonia after a few days, this indicates that the beneficial bacteria are low or nonexistent. Use products that introduce good bacteria to the aquarium which will take about a week to be fully effective. Using zeolite chips to absorb ammonia in a separate box filter for a week. This is the best solution until the bacteria can take effect.

Conclusion

Many people have a love for aquariums, but an aquarium is not just for fish. You don't need saltwater to get a beautiful and interesting aquatic scene. Freshwater aquariums are gaining in popularity, and you can find them in every price range. So if you love watching the different creatures of the water world, it's time to build your own freshwater aquarium!

What are some of the benefits of freshwater?

All-Natural – Unlike saltwater tanks that require regular changing of water and adding chemicals, fresh water tanks only need regular maintenance with ordinary tap water or well-filtered bottled/spring water.

No Corrosion – Since there is no salt, there is no corrosion to the equipment. Rusting will not be an issue as with saltwater tanks.

Ease of Maintenance – A freshwater tank offers an easier maintenance schedule that is more consistent than that of a saltwater tank. A freshwater tank needs to have its water changed every couple of weeks or so, while a saltwater tank needs changing every other day!

Since fresh water fish are not encased in a protective environment, they need the added benefits freshwater provides:

Background – Freshwater aquariums work well with almost any kind of background you can think of from rocks to artificial pine trees. You can use full backgrounds that hide the fish tank equipment, or you can use simple backdrops that are just for show.

Filter – Freshwater tanks need a filter that is of good quality and strong enough to keep up with the waste material from the fish and plants. Look for filters that have attachments to attach an external filter in case there is too much waste material.

Lighting – Freshwater tanks need lighting, too. They don't need as much as saltwater tanks, but they still do require at least two hours of light per day or they will become stressed and possibly sick. You'll want some strong lighting to help develop healthy leaves on your live plants.

Lighting can be as simple as two bright desk lamps or as elaborate as a complete light fixture. The best thing about freshwater aquariums is that you can try for natural lighting and avoid the harmful UV rays of a regular light.

No Salt – Freshwater tanks require no salt, which means you don't have to worry about the fish being exposed to too much salt. Also, since there is no salt, there won't be any corrosion on your equipment.

Plants – Most freshwater fish like to swim over plants, especially live plants. It gives them hiding places and helps keep the water clean at the same time. You'll want live plants in your tank not only for the fish but also for your enjoyment.

Plants are not required, but if you do have plants in your tank, you'll want to monitor the water parameters. Live plants require more care than a fake plant, and you don't want to catch anything on them!

Fish Health – One of the best benefits of freshwater aquariums is that they keep fish company. You'll find many people keep goldfish and other small fish as pets that love nothing more than swimming around in a tank full of other fish. You might even find yourself keeping algae eating aquarium shrimps or crayfish which are popular for their ability to clear out unwanted algae from all types of aquariums.

Freshwater Aquarium Supplies

You don't need a lot of fancy equipment just to get started. Here are the basic supplies for a freshwater tank:

Aquarium – Just like setting up any fish tank, you'll want to get one that is big enough for the fish and plants. You'll also want one that is tall so that you can grow live aquatic plants without them being covered by the water. Gather all of your fish and plants first and then pick out an aquarium that will work for them all.

Check the kit that comes with the tank as it sometimes has everything you need to get started. The kits usually come with fish food, test kits, and other supplies. Some aquariums even have starter chemicals that you just add to the water for a few days until you're ready to start adding your fish and plants.

Substrate – In addition to gravel or sand for your tank bottom, most people choose aquascape products like marbles or slate pieces for a more natural look. You'll also want a non-toxic substrate cover so that your fish don't get sick from chemicals used in some substrates.

Aquarium Stand – Since you will need a sturdy stand to hold your tank, make sure that it's as stable as possible. Some aquariums can be very heavy, especially if they have a lot of water and/or plants in them. Choose your stand carefully to support the weight and size of your fish tank.

Aquarium Hood – Hoods can help filter the air in your tank, keeping harmful ammonia and nitrite from escaping. They also allow you to keep the lights on in the tank without hurting your fish.

You can choose a smaller or larger hood depending on how much light you want to have in your tank. You will need to have a heater for your hood if it's outside or if you plan to keep lights on overnight. It is possible to use aquarium lights, but they are not as efficient as regular aquarium equipment.

CPSIA information can be obtained
at www.ICGtesting.com
Printed in the USA
LVHW082140250521
688474LV00002B/284